Disclaimer

The publisher of this book is by no way associated with the National Institute of Standards and Technology (NIST). The NIST did not publish this book. It was published by 50 page publications under the public domain license.

50 Page Publications.

Book Title: Rethinking Egress: A Vision for the Future

Book Author: Jason D. Averill; Richard D. Peacock;

Book Abstract: New technologies and research are redefining the state-of-the-art in building evacuation. The time is right to rethink the entire infrastructure of egress from buildings in light new opportunities to address the economic and life-safety issues. Approximately 40 experts from a variety of disciplinary background assembled in Warrenton, VA from April 1-3, 2008 in order to consider building evacuation, starting with a blank sheet of paper. Structured around the principles of Value-Focused Thinking (a text authored by workshop moderator Ralph Keeney), the participants were encouraged to consider values, objectives, alternatives, and metrics. This process combined the benefits of free-thinking brainstorming with a formalism which encouraged evaluation of the potential for new ideas. By the conclusion of the third day, over 400 ideas had been developed, along with metrics for future evaluation of the ideas.

Citation: NIST TN - 1647

Keywords: Fire; safety; high-rise buildings; egress; evacuation

NIST Technical Note ZZZZ

RETHINKING EGRESS:
A VISION FOR THE FUTURE
Workshop Proceedings
April 1-3, 2008, Warrenton, Virginia

Jason D. Averill and Richard D. Peacock

Building and Fire Research Laboratory

Ralph Keeney

Duke University

Carita Tanner and Claret Heider

National Institute of Building Sciences

September 2009

U.S. Department of Commerce

Carlos M. Gutierrez, Secretary

National Institute of Standards and Technology

Patrick D. Gallagher, Deputy Director

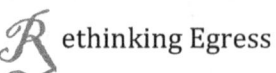

Certain commercial entities, equipment, or materials may be identified in this document in order to describe an experimental procedure or concept adequately. Such identification is not intended to imply recommendation or endorsement by the National Institute of Standards and Technology, nor is it intended to imply that the entities, materials, or equipment are necessarily the best available for the purpose.

National Institute of Standards and Technology Technical Note ZZZZ

Natl. Inst. Stand. Technol. Tech.Note ZZZZ, X pages (September 2009)

CODEN:

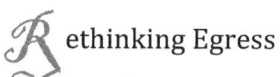

ABSTRACT

New technologies and research are redefining the state-of-the-art in building evacuation. The time is right to rethink the entire infrastructure of egress from buildings in light new opportunities to address the economic and life-safety issues. Approximately 40 experts from a variety of disciplinary background assembled in Warrenton, VA from April 1-3, 2008 in order to consider building evacuation, starting with a blank sheet of paper.

Structured around the principles of "Value-Focused Thinking" (a text authored by workshop moderator Ralph W. Keeney), the participants were encouraged to consider values, objectives, alternatives, and metrics. This process combined the benefits of free-thinking brainstorming with a formalism which encouraged evaluation of the potential for new ideas. By the conclusion of the third day, over 400 ideas had been developed, along with metrics for future evaluation of the ideas.

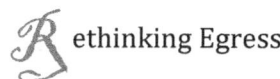

CONTENTS

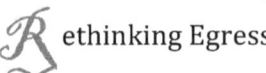

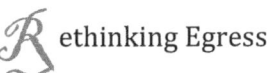

BACKGROUND

Following the Federal Investigation of the Collapse of the World Trade Center and subsequent report, the National Institute of Standards and Technology (NIST) partnered with with the National Institute of Building Sciences and its Multihazard Mitigation Council (MMC) to implement recommendations. A group of recommendations focused on proving reliable egress for occupants from high-rise buildings. Specifically, "building evacuation systems should be improved to include: system designs that facilitate safe and rapid egress; methods for ensuring clear and timely emergency communications to occupants; better occupant preparedness for evacuation during emergencies; and incorporation of appropriate egress technology.

During its deliberations, the MMC committee reviewing the report recommendations concluded that the use of elevators for evacuation and firefighter access in high-rise buildings had the potential to revolutionize the process of evacuation to such a significant degree that a complete reevaluation of egress systems was warranted. A workshop of multi-disciplinary experts in the field would provide a forum for discussion and innovative insights. In collaboration with NIST staff, a small number of technical experts were identified to assist with planning the workshop, shown on the left of this page.

Workshop planning sessions were held in-person and by conference call, throughout 2007. Early in the planning, the value of a professional moderator with expertise in structured brainstorming was identified. Prof. Ralph Keeney (Duke University) was retained in order to help organize the workshop around the framework described in his seminal text, "Value-Focused Thinking: A Path to Creative Decisionmaking."* Subsequently, the workshop was scheduled for April 1-3, 2008, at the Airlie Conference Center in Warrenton, Virginia.

Rethinking Egress Planning

- Najib Abboud of Weidlinger Associates

- Carl Galioto of Skidmore, Owings & Merrill, LLP

- Clas Jacobson of United Technologies

- Dennis Mileti, Professor Emeritus of the University of Colorado

- Russ Sanders of the National Fire Protection Association

Rethinking Egress Moderator

- Prof. Ralph Keeney of Duke University

---

* Keeney, Ralph L. "Value-Focused Thinking." Harvard University Press, Cambridge, MA. 1992.

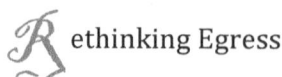

> There is nothing more difficult to take in hand, more perilous to conduct, or more uncertain in its success, than to take the lead in the introduction of a new order of things.
>
>
> "The Prince" - 1513
>
> Niccolò Machiavelli
> (1469 - 1527)

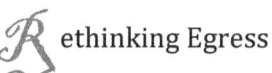
ethinking Egress

WORKSHOP FORMAT

OUTLINE AND PROCESS OF THE WORKSHOP

This section describes what was done in the workshop and why. The agenda for the workshop is shown in Appendix B. The substance of the first morning was five presentations from different perspectives on the problem of egress. In the first afternoon, individuals worked alone using a variety of techniques to generate egress alternatives and a set of objectives that they hoped would be achieved by such alternatives. This provided a foundation for the next two days.

In the morning of day 2, participants met in groups of approximately seven to create additional alternatives based on their collective thoughts. Then the entire workshop discussed some of these alternatives and embellished them. In the second afternoon, the individual groups considered in detail two processes: effective egress and fire fighting. Three groups identified all of the steps that would be necessary for an effective egress from a very large building, and one group listed all the steps necessary for fighting a fire in a very large building. These steps were then used to generate still more potential alternatives to facilitate egress.

The third morning of the workshop included a preliminary evaluation of a subset of the alternatives that fell into each of the categories of alternatives that had been previously developed. One purpose was to find out the types of alternatives that seemed most useful that had a high feasibility of being practical in the near future. A second purpose was to identify those categories of alternatives that were particularly creative, and hence perhaps more appropriate for further in-depth creative activities to develop additional alternatives.

The reason the workshop was organized as described above was based on three premises: (1) all thinking is done in individual's minds, as there is no collective mind, (2) in discussions, individuals tend to narrow their thinking to focus on the discussed content, and (3) individuals working in a group can expand on the ideas of others. Hence, we wanted to stimulate thinking by the individuals to get all of the relevant ideas from each individual first, and then expand these with personal interaction. Thus, in the morning of the first day, we initially ask individuals to write everything down that they thought would be relevant to the workshop, so that they would be less concerned about exactly when their points of view would be discussed and more willing to create new ideas. The afternoon focused on explicitly articulating what was in individual's mind to provide a large number of sources for generating alternatives. These generating activities provided a large number of alternatives, and subsequently each individual was better prepared to contribute to the group discussions and group creation of additional alternatives on the second day.

To understand the benefit from the set of objectives for egress, it is necessary to organize the union of all the objectives identified by individuals. Experience has shown that it is very difficult for an individual to come up with all of the objectives for a given situation. Thus, although many individuals naturally identify some of the same objectives, each individual typically identifies some objectives that were not initially articulated by other individuals, but that are relevant to them. In fact, on the second morning, the entire workshop basically accepted the set the objectives organized on the first evening as appropriate for the problem.

8

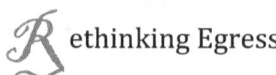

THE OBJECTIVES OF EMERGENCY EGRESS.

An objective is a statement of something that one desires to achieve in a given decision context. In our situation, the decision context is emergency egress from large buildings. It is important to distinguish between two types of objectives: fundamental objectives and means objectives. The fundamental objectives characterize the essential reasons for interest in the decision situation. A means objective is of interest because achieving it can contribute to achieving the fundamental objectives.

To identify the objectives for a given decision situation is not easy. One begins with the set of values appropriate to the decision, where values are defined as anything that one cares about concerning the decision.

The concepts and techniques of value-focused thinking (Keeney, 1992) were used in this workshop to stimulate the identification and creation of alternatives to facilitate egress from large buildings. First, all the values of each of the individuals were identified. These values were then used to specify a clear and coherent set of objectives for emergency egress. Each of the objectives identified by an individual was then used by that individual to create alternatives that could better achieve those objectives. Subsequently, individuals worked in groups to identify additional alternatives through their collective creative processes. You might summarize the entire procedure as one of value-focused brainstorming.

In identifying the list of values that provide a basis for objectives, we began by first asking each individual to list all values they felt were appropriate for the decision. However, experience suggests that almost everyone ends up with an incomplete list, so there are a number of techniques to stimulate thinking about additional values. The techniques that we used in the workshop are given in Appendix D.

Once a set of values was listed by an individual, the individuals were asked to restate each value as an objective. An objective can be stated using a verb and an object, such as maximize safety, minimize lives saved, or avoid property damage. With this common format for stating objectives, they are much easier to understand and much easier to structure.

After each of the individuals converted each of their values to objectives, we organized all of those objectives on the first evening of the workshop. Taking over 400 objectives from the individuals, we created the means-ends objectives network in Figure 1. This network includes the fundamental objectives and the means objectives for decisions concerning emergency egress. The basic information, meaning all of the individual's objectives, used to create the means-ends objectives network is presented in the workshop findings. This list deleted multiple statements of the same objectives and organized the objectives into categories.

It is important to point out that each objective should concern only one object of importance. Value tradeoffs among the achievement of different objectives are obviously critical in evaluating or choosing alternatives, but such value tradeoffs are not desirable at the stage of listing and organizing objectives. As an example, one means objective is to minimize egress time and one fundamental objective is to minimize injuries to individuals. There certainly would be cases were rushing the egress to reduce the egress time to a minimum could result in additional injuries to those evacuating the building. This would not be preferred to a slower egress that did not cause those injuries. This preference would naturally be addressed in dealing with the value tradeoffs

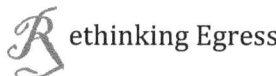

among objectives and with the consequences that the means objectives have on the fundamental objectives. Those tradeoffs must be recognized, but we do not need to state a means objective such as to "reduce egress times while maintaining safety", as this would combine two objectives and the implied causal effects and value tradeoffs into that one objective. To do so mixes up the notions of what are the objectives and what are the relationships, both causal effects and value tradeoffs, among them.

THE ALTERNATIVES GENERATED FOR EMERGENCY EGRESS

An alternative is defined as any action that is totally under the control of a decision maker that may influence the degree to which at least one of the objectives of emergency egress is achieved. The decision maker can decide to implement it or decide not to implement it. Every alternative should at least contribute positively to achieving one of the means objectives or one of the fundamental objectives. If this did not happen, then of course there would be no reason to consider it.

During the workshop, there was a discussion about whether the term alternative should be something different including option, tactic, strategy, choice and so forth. In this document, we mean that the alternatives are the aggregation of any of those similar ideas, where similar means something under the control of decision makers that can be done that would influence egress.

It is important to recognize that most of the alternatives would not be what could be considered to be a complete alternative of everything that would be done. One might rather think of these as elements of alternatives. Specifically, one could do something that would facilitate movement of disabled people in stairways in a particular building. This does not mean that would be the only thing that would be done to facilitate emergency egress. Other alternatives that might reduce the size of a fire once it had started in a building or that would better communicate information to people about how to egress in the event of an emergency could also be implemented. Thus, it is appropriate to consider many of the alternatives generated in this workshop as elements of what might be consider a complete package of alternatives that could be implemented.

The complete list of alternatives developed at the workshop is given in the workshop Findings. In addition, this section presents are the categories of alternatives that the workshop team subsequently developed on the second evening to facilitate communication and organization of the alternatives. This organization was also helpful for a rough appraisal of some alternatives as discussed in the next section.

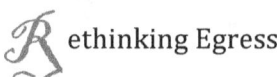

As a basis to stimulate thinking about alternatives and partly as a trial to examine evaluation procedures, each of the participants were asked to evaluate about 40 alternatives selected by the workshop staff on the last day of the workshop. We present the results, but it is important that readers understand that the results are very preliminary for several reasons. First, the alternatives were not necessarily well-defined, as they were described by only a few words. To do this better, each of the alternatives should be a much more clearly defined. Second, information was not available about details of the alternatives, such as how it might be implemented. Third, different participants naturally had their expertise in different aspects of egress and may not have had a complete understanding of other aspects. And the knowledge of some participants about certain aspects of an alternative was not available to other participants. Recognizing these caveats, we still felt it was useful to go through the preliminary evaluation, shown in the workshop findings.

That appendix gives a very quick appraisal of the summary comments of the various individuals. The alternatives were evaluated by nine groups of three individuals. They were evaluated on three criteria: quality, feasibility, and creativity. Quality was defined as how good the alternative is if it could be effectively implemented. Feasibility was defined as the chance that the alternative could be effectively implemented within 10 years. Creativity concerned how well known the idea was versus whether it was a completely new idea. Quality and creativity were evaluated on metrics ranging from 1-9. Feasibility was described in terms of the judgment, using percentage, that the alternative could be effectively implemented in 10 years. An advantage of this feasibility scale is that it aggregates both economic and technological feasibility, but that is also a shortcoming with a quick appraisal because it is not easy to aggregate intuitively those two types of concerns.

The evaluation represents very quick simplistic aggregation done during a break at the workshop. When there was general agreement of the nine evaluation teams, roughly meaning that the dispersion of estimates were less than a third of the range of the corresponding measure for at least eight of the teams, we used the mean response. When there was more dispersion among responses, we included the range for feasibility and two estimates for each of the 1-9 scales that represent mean responses for the low and the high sets of evaluations with no characterization of how many were in each of those groups.

From observing the teams in the process of discussing and evaluating the alternatives, it appeared that it would be worthwhile to develop a more substantial and clearer evaluation form and process. This may provide insights about the low-hanging-fruit, meaning the alternatives that almost everybody would agree on in terms of overall effectiveness and on how quickly they can be implemented to improve egress. This evaluation would also try to identify some types of the alternatives where very creative ideas might have a big impact in the future.

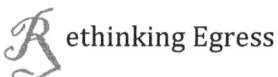

The set of objectives provide a sound basis for creating alternatives and for subsequently evaluating alternatives for emergency egress. It also provides a basis for modification and improvement. Given the way the set of objectives was obtained as a collection from numerous knowledgeable individuals, we would expect that the scope of these objective would incorporate most of the concerns one would have with emergency egress. However, there are likely some concerns missing and many of the concerns that are explicitly included might be focused more clearly with additional thinking.

The set of alternatives for emergency egress documents a large number of potential actions that one can consider. This should provide a foundation for any additions that might be made in the future. It seems reasonable that there would be thousands of potential alternatives that could facilitate emergency egress from buildings. The set of alternatives documented here, and their categories, begin to define the full scope for most of the types of alternatives that might be feasible within a 10-20 year period. We believe that additional types of alternatives can likely be identified and within both new and existing categories, additional focus in creating alternatives would be useful.

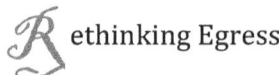

WORKSHOP SESSION SUMMARY

The principal objective of this highly-interactive, free-thinking workshop was to facilitate a reassessment and definition of egress objectives and to develop of a vision for the future of the means of egress, especially in tall buildings. The agenda for the workshop is shown in Appendix B.

APRIL 1, 2008: DAY 1 OF THE WORKSHOP

AVERILL: WELCOME

Jason Averill of NIST opened the workshop by welcoming everyone present, along with the wider audience participating via webcast. Those watching the webcast were able to submit questions to be addressed during the workshop deliberations. First, Averill highlighted some of the constraints of current egress systems and practices in high rise buildings. A concept as simple as building stairs is poorly understood: for example, how much benefit is conveyed to the occupants by an incremental change in stair width? Given the substantial up-front, life-cycle, and opportunity costs of stairwells, a stronger technical basis for performance should be established.

Next, he urged participants to consider the elevators. Since everyone enters a building by elevators, and we know that people tend to exit the way that they entered, shouldn't we use elevators to exit most tall buildings? As buildings are designed to be taller than ever, the impetus to improve our understanding and performance of egress system is magnified.

Averill contended that the current state of egress requirements reflects three realities: lack of basic research, lack of concentrated effort directed at the problem, and regulatory inertia. This workshop will be a first step towards an egress system which is reliable, robust, cost effective, and well-understood.

Averill expressed that the participants available in person reflected the continuum of disciplines which can contribute to the safety of people in buildings during egress. He also emphasized that the mission of the highly interactive, free-thinking workshop now convened was to explore emergency egress in buildings considering the interaction of the physical and social environments. Averill articulated the ground rules for the workshop:

• Everything discussed is part of the public domain.

• The discussion is to be positive because an atmosphere of creativity, free thought, and sharing yields better results.

• Participation by everyone is of the utmost importance.

• Participants should be open to the process of structured brainstorming.

After reviewing the logistics for on-site participants and introducing the experts who assisted with the planning of the workshop, Mr. Averill turned over the floor to the moderator, Professor Ralph Keeney.

KEENEY: WORKSHOP PROCESS

Prof. Keeney briefly described his background, and reviewed the purpose of the workshop. The principal resource available for the success of the workshop was the vast experience of the assembled experts. Therefore, the purpose of the moderator is to enable the transfer of the vast experiences in the minds of the participants onto paper. After briefly reviewing the agenda, Prof. Keeney turned the floor over to the presenters. The remainder of the morning was devoted to presentations. The purpose of the presentations was to establish a foundation of common knowledge that would enable all of the participants to better apply their knowledge in the context of the other discipline. Presentations were delivered by Russ Sanders, Clas Jacobsen, Carl Galioto, Dennis Mileti, Najib Abboud, and Bud Nelson. Note that the PowerPoint presentations of can be found in Appendix B.

SANDERS: FIRE DEPARTMENT OPERATIONS IN HIGH-RISE BUILDINGS

Russ Sanders reviewed fire department operations with respect to building occupant and firefighter safety. As a context for operations, Sanders discussed risk assessment and mitigation. Risks for the fire service responding to a building fire tend to increase with the height of the building for several reasons:

- Most aerial apparatus can reach only 8 stories. This removes the option of rescue by fire department ladder above 8 stories.
- Master streams are rarely effective and therefore, the fire service is forced to rely on offensive interior operations, which tend to be the most dangerous operations.
- Lead times to get people and equipment (tools and air bottles, e.g.) high in a building can be significant.
- The taller the building, the larger the number of people who can be caught above the fire.

Sanders then reviewed the process of fire scene size-up, including determination of building type, size, time of day, location of fire, presence of built-in protection systems (sprinklers in-particular).

Once on-scene, the incident commander (IC) has three basic priorities upon arrival on the scene of a building fire:

- life safety,
- extinguishment, and
- property protection.

These are not independent, but interdependent concerns. In order to address these concerns, the fire service must address complex and evolving challenges, including ventilation within buildings, fire suppression, situation awareness, and the interface between fire service and occupants in the stairwells. For example, one of the most complex aspects of high-rise firefighting is ventilation. While the fire service puts out a fire with water, the fire service controls the fire by controlling the ventilation. Proper ventilation can dramatically reduce the risks to the occupants and firefighters, while improper ventilation can result in civilian or firefighter deaths.

The only way for the fire service to verify that a building has been evacuated is to conduct both a primary and secondary search. This requires that the fire service maintain control and integrity

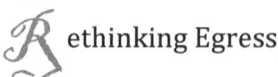

(free of smoke and heat) of the stairwells. Pre-planning and knowledge of the characteristics of the building when the fire department arrives on-scene is a significant advantage. This requires active participation of the building management who are most knowledgeable about the specifics of the fire safety systems.

JACOBSEN: LEVERAGING INFORMATION IN BUILDING EVACUATION

Clas Jacobsen reviewed the role of information technology in building evacuation. He told us that we need to manage the flow of mass, energy, and bits. Mass is the flow of people through the building, energy is the fire, while bits is the data and information; in modern buildings, the bits are growing dramatically over time. The challenges going forward are to use computation and communication to enhance egress performance.

We have dynamically evolving information flows. In order to harness this information, the building community needs to consider three basic questions: who pays, who plays and who decides? An important testbed for understanding the evolution of these information systems is modeling. Modeling makes precise what the underlying problems in the system are, while also providing uniformity.

There are many problems with integrating all of this information into a single systems approach.

- Systems issues are multiscale and multiphysics.
- Uncertainty propagation in the system can overwhelm the utility of the output. However, one approach is to consider the use of reduced-order models
- Codes and regulations need to keep up with "cyberphysical" systems: IT controlling physical systems.
- The information should be available in multiple locations: off-site (en-route for firefighters) and on-site (through a fire panel).
- Information flows are diffuse and lack organization. Opportunities for analysis and presentation to occupants and firefighters should be considered, such as decision support tools.
- The human-machine interface requires development.
- Current generation predictive models currently are computationally expensive. New techniques in reductive methods should be considered in order to produce answers on the time scale required for actionable results.
- The interfaces between components are where many problems emerge. The systems are dynamic; in particular, the time scales do not always align.

GALIOTO: CHALLENGES FACING ARCHITECTS FOR SUPER-TALL BUILDINGS

Carl Galioto provided a review of the state-of-the art of super-tall building design using examples from the tallest buildings currently under construction. Many of the occupants who inhabit the highest portions of super-tall buildings (observation decks, restaurants, hotels, assembly spaces, e.g.) also tend to be unfamiliar with the egress systems in the building.

Growing economies are building towers taller than ever before. Along with the height of the building come increasing demands for security at several levels: perimeter, enclosure, core, and emergency access core. These requirements often conflict with the goals of building egress and

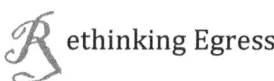

emergency services ingress. One solution which included use of pressurization systems and protected elevators currently utilized in high-rise design was reviewed.

Galioto compared the increased risk of designing high-rise buildings to the increased risk of ships out at sea. Those ships require life boats which are able to protect passengers from an adverse event on the ship. Super-high-rise buildings require innovative egress solutions which are comparable to the risks inherent to the building under consideration.

MILETI: THE SOCIAL SCIENCE OF BUILDING EVACUATION

Dennis Mileti summarized the state-of-the-art knowledge base from psychology, social psychology, sociology, and human factors as it related to building evacuation or other appropriate protective actions. The research base for the talk extends over 60 years and was compiled by Kuligowski and Mileti.

The factors which influence human behaviour are well-known. By consolidating and implementing this body of knowledge, egress performance can be improved. Key findings include:

- Panic can happen, but is rare and words cannot prevent or cause panic. However, building design can influence panic: e.g., doors that swing outwards and sufficiency of egress paths.
- Building evacuation is not a continuous process. It happens in "lumps and bumps."
- Alarms interrupt ongoing life. It simply sends people in search of additional information. People don't remember indicators, unless it is drilled into people, relentlessly. People are hard-wired to think they are safe, that disasters happen to other people, and that it is probably a false-alarm. To alert people, it should be intrusive as possible.
- Conflicting information abounds. Formal alerts compete with informal alerts
- Seconds of delay can cost lives in a fire emergency, but people are wired to perceive that they are safe. Until they confirm the risk through milling or external cues, occupants will delay.
- When instructions are delivered to occupants, several aspects of communication are well established in the literature:
  - Channel: the more the better. If you want people to take an action, warn them through as many channels as possible.
  - Frequency: the same message should be repeated as many times as possible, Madison Avenue has known for years that people will not remember something until you repeat it ten times.
  - Content: the single most important message that can be delivered to an occupant is what you want them to do. Don't talk about the fire or the earthquake, but give them specific, actionable advice.
  - Style: keep it simple, precise, authoritative, accurate, and avoid jargon. Be internally consistent.
  - Source: tell them who is speaking. The most authoritative source is different depending on the situation, but surveys indicate that firefighters are the most authoritative source in the United States (but for only 35% of the population).
- Cues are critical. Fire trucks, smoke, flames, and floor wardens who are leading an evacuation are all strong cues. Meanings may be different, however, depending on the filters that occupants bring to the event. Filters include education, age, gender, race, language, associations with co-workers, parental status, social isolation, experience, e.g.

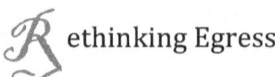

- Information needs do not cease when the alarm is sounded. Once the occupants are started, they still require additional guidance.

ABBOUD: STRUCTURAL ENGINEERS, FIRE, AND EGRESS

Najib Abboud discussed how the patterns of increased urbanization are driving the trend towards taller buildings. The trend is being set on a worldwide stage, and is currently focused on multi-use skyscrapers (vertical cities). The driving force behind the multi-use prototype is risk management: if one real estate sector is down, the other use groups can maintain a revenue stream for the owner. This presents challenges for a single building core to serve all of the different occupancies, both structurally and from an egress perspective. With that background, Abboud then summarized three challenges for the structural engineering community: public policy, integration of fire and structural engineering, and materials innovations in response to new threats.

The first challenge is establishment of public policy. Standardization sets a level playing field from an economic perspective. Without standards, few projects will break the trend. What is safe? What is life safety? How long do you need the building to stand? These are the public policy questions that codes and standards need to specify. The technical challenges are achievable once the public policy is established.

The second challenge is to integrate fire engineering with structural engineering. The two disciplines are as unintegrated as can be. The only way that structural engineers can think of fire is as a load. Conversely, the information needs to be provided to the structural engineer in those terms: intensity, duration, location, and statistical probability. Hurricanes and earthquakes are reduced to return frequency, and fire needs to be considered similarly.

The third challenge is to adopt lightweight, flexible materials which can resist a variety of loads, including blast, fire, and vandalism. Traditional materials are either too heavy to build tall buildings with or too light to resist the variety of loads imposed. New materials such as one-inch, ultra-high strength concrete are being developed to ensure integrity of the core, including stairwell enclosures which can preserve exit paths for evacuating occupants.

NELSON: THE LEGACY OF THE 1971 AIRLIE HOUSE HIGH-RISE LIFE SAFETY CONFERENCE

To conclude the invited presentations portion of the workshop, Bud Nelson shared his experience with the 1971 Airlie House Conference of High Rise Life Safety, which served as an inspiration for th current workshop. The objective of the 1971 conference was to bring independent disciplines together and develop the concept of a systems approach to fire in high-rise buildings. The first element of the systems approach was the threat from fire. The second element was force modification (sprinklers, compartmentation, etc). The third element was information and guidance: response of the exposed and protection of the exposed. All of these systems concepts can be translated directly to an analysis of the egress system. However, much additional research will be required in order to prevent 'garbage in, gospel out.'

The 1971 workshop had tremendous impact on the design of today's high rise buildings, including the advent of deterministic modelling. Concepts of fire safety evaluation system, performance based design, and systemic building design can trace many of their roots right back to the Airlie House

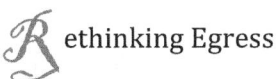

Conference Center. Indeed, a workshop can have tremendous impact. But only if the participants carry forward the ideas.

ALTERNATIVES AND OBJECTIVES

During the afternoon session, participants worked independently to identify egress alternatives and objectives for the workshop. Using forms distributed by Prof.. Keeney, participants were asked to put their thoughts on paper and then to discuss these thoughts as a mechanism for generating more ideas. (The forms used in this and the forms used in subsequent workshop exercises are included with the workshop agenda in Appendix A.) Once completed, the forms then were collected with the promise that they would be copied and returned and that the thoughts and ideas written on the forms would later be summarized by NIST staff and the summary distributed so that it could serve as the foundation for later discussions. Participants devoted the remainder of the afternoon session to:

- Listing alternatives that may enhance egress,

- Listing objectives that may enhance egress,

- Listing values (what the participants care about regarding egress),

- Expanding the list of values,

- Convert each of the values to objectives, and

- Creating alternatives or elements of alternatives using each objective.

Prof. Keeeney pointed out that listing both alternatives and objectives forces a decision to be made but that having both options makes for a better decision. Objectives define decisions and alternatives contribute to that. He noted that the key of these exercises is to think outside the box.

During the evening of the first day, the NIST staff and Prof. Keeney synthesized the results and developed a comprehensive list of objectives. The list can be found in the "Workshop Findings" section of this report.

Prof. Keeney greeted everyone and explained the agenda for the day and how the results from the previous day were synthesized. He referred to a whiteboard version of the following:

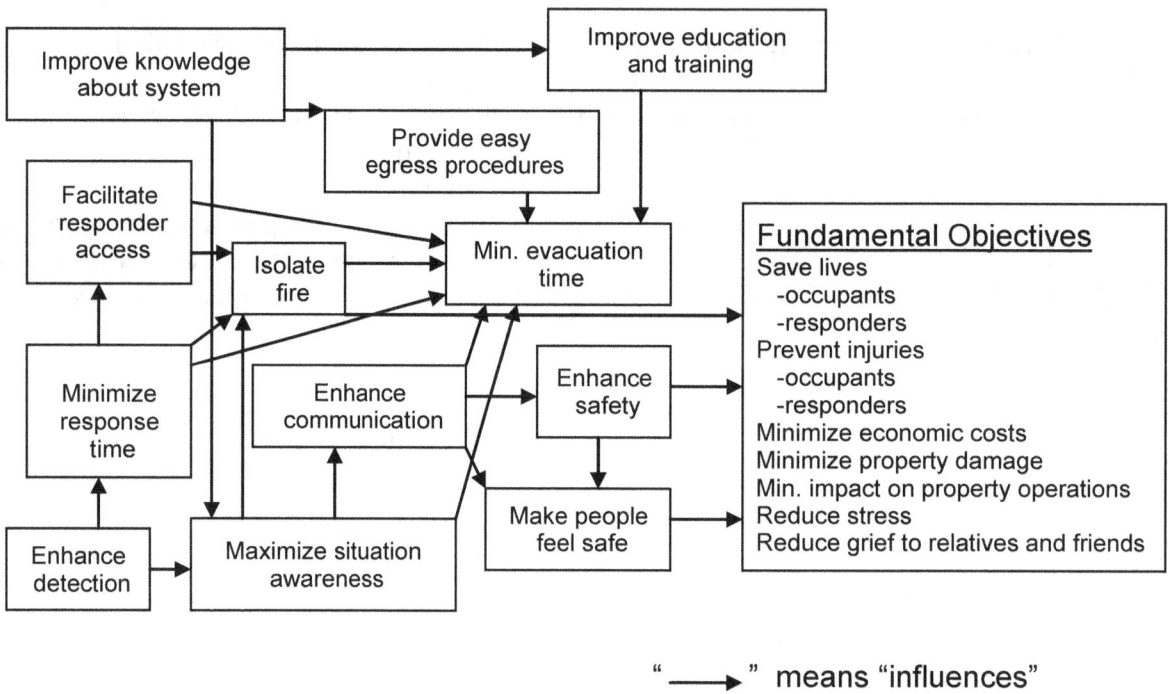

"——▶" means "influences"

FIGURE 1: MEANS-ENDS OBJECTIVES NETWORK FOR EMERGENCY EGRESS

Prof. Keeney noted that the fundamental objectives listed by the workshop participants were very similar and could be summarized by the following objectives:

- Save lives of occupants (many classes) and of firefighters/responders

- Minimize property damage

- Minimize impact on property operations

- Minimize economic costs

- Reduce stress

Mr. Averill pointed out that they arrived at these fundamental objectives by grouping participant responses into general categories and then arranging them. Prof. Keeney said the most important objective obviously is life safety. He added that there are ways to be safe that do not involve leaving a building but that the focus of this workshop was egress from high-rise buildings.

Prof. Keeney said that the alternatives listed on the forms could be applied to all fundamental objectives as well as the main one, saving lives. He also mentioned that among the responses there were some objectives that had crept in among the alternatives as well as some alternatives among

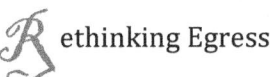

the objectives. He stressed that the alternatives are the things that can be controlled. The remainder of the morning session was devoted to working group discussions to expand the alternatives.

During the afternoon, the working groups focused on effective egress and fire fighting. The result was a detailed description of the steps involved in effective egress from a very tall building, and additional alternatives for facilitating egress were identified. The deliberations of each of the working groups are summarized below.

•   Group 1 - Members of this group were not sure if they were correctly differentiating between objectives and alternatives. The group generally agreed that minimizing evacuation time was the highest priority and refined this objective to read: "Timely evacuation with minimal harm to occupants." A number of safety criteria were mentioned – for example, minimizing fatalities and injuries and maintaining building security. High on the priority list was maximizing the comfort level of occupants, be it physical or psychological. Lower on the priority list were cost and robustness of the solution. The group also addressed the various scenarios that can occur and noted that any solution must address the full range of scenarios. Components and routing systems would need to be used in as wide a range of situations as possible and this will have benefits in terms of developing occupant familiarity with the components/systems as well as in terms of maintenance and cost if the components/systems can be used in nonemergency situations. With respect to stairwells, it was noted that occupants tend to avoid them unless they use them regularly Appropriate communication/information systems are very important and how they will be used can be treated during training exercises but their use can also be addressed during the design process. If the design is intuitive and obvious, it will encourage people to use certain routes and become intimately familiar with them. In those situations where occupants are likely to be strangers to the building, staff must be very well trained to manage the event.

•   Group 2 - The group decided that alternatives actually are the tactics that can be used to achieve objectives. Expanded and increased communication was deemed to be very important both prior and during an event but also after the event. Two-way communication between the fire responders and the building occupants was identified as being of special importance. Increased use of elevators for emergency egress was discussed as were potential trade-offs that might be made in design. Group members held varying opinions on elevator use during emergencies and it was concluded that research is still needed. Evacuation devices also were also discussed and additional research suggested. The need for increased planning for all types of threats was addressed as was the need for measurable performance criteria and improved sprinkler reliability. It was suggested that enhanced situational awareness be added as to the list of objectives.

•   Group 3 - Group 3 discussed many of the same issues as Group 2 but emphasized that evacuation is the crux of the egress problem. For example, under what scenarios do buildings have to be completely evacuated and do they really need to be evacuated. The group concluded that one general way to address this issue is to develop a process for creating threat scenarios and establishing performance metrics for providing alternatives acceptable to stakeholders. The scenarios could involve a security breach, a massive fire, or a CBR attack. By analyzing these scenarios, we can provide building designers with the tools they need to meet the performance objectives for the building. The next step would be to develop a standardized but flexible system for communicating instructions to occupants and emergency responders using conventional communications devices and technology. Also discussed were the use of sky-bridges between

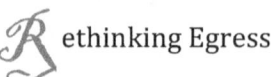

buildings, the use of external evacuation systems, maintaining incident command from a remote location, having stairwells discharge directly outside, and the need for improved sprinkler and monitoring systems.

• Group 4 - This group concentrated on alternatives for making people feel safer, and the general consensus was that the best way to do that is through increased use of technology. The systems currently used buildings require enhancement – that is, buildings need to "see," "hear," and "talk' to us. The group concluded that the best way to evacuate a building is through the use of elevators (even though simple egress methods such as stairwells would still need to be present); therefore, all aspects of making this possible should be explored. Zone-floors or safe areas and enhanced fire resistance for high-rise buildings also were discussed. Several group member indicated that the building codes permit too many trade-offs when sprinkler systems are provided.

Considerable discussion was devoted to the sky bridge concept and the idea of buildings sharing resources and equipment. Opinions varied on the ideas. It was noted that the use of sky bridges is being explored in Australia but that it is a massive undertaking that requires a totally new way of thinking and modified building codes. Others indicated that the concept is a difficult one to use and should be considered only on a case by case basis. Still others suggested that egress is only one benefit of sky bridges; they can be used to create an urban network with vertical links within buildings and people become familiar with their evacuation routes because those are the routes they use on a daily basis. One group member mentioned the 1969 book, Urban Design Manhattan, which explored the idea of interconnecting buildings on many levels but did not specifically address emergency egress. A proposal for redevelopment of the World Trade Center complex called for the use of sky bridges but it was unsuccessful; it would have been a very good example of how to develop a large site into one complex of functional, integrated buildings. Technical design issues involving such things as different response frequencies were mentioned as were the legal issues (not simply a contractual issue but also an entitlement issue in real estate law). It was noted that elevators account for about 15 percent of a tall building's energy usage and that sky bridges linking lower floors could reduce elevator use and, consequently, energy use and cost.

During discussion of the egress process and the fire fighter process, it was concluded that both involve the following:

• Determining that an event has occurred,

• Verifying that the event is real or false,

• Notifying people that an event is taking place and action is needed, and

• People receiving this information and determining the correct action to take.

Differences in function – high rise buildings of mixed occupancy and use, assembly buildings, and general use buildings – were discussed in terms of the impact building use has on egress system requirements.

During the afternoon session, the group reconvened to hear and debate the results of the working group discussions. Mr. Averill suggested the each group provide a summary of their discussion and identify the steps taken in the process as well as alternatives discussed and other ideas addressed for a significant amount of time. Summaries of the working group discussions appear below.

•   Group 1 - The process discussed by this group was phased, rather than linear in that they discussed the overall event timeline and the participant experience timeline in terms of the detection phase (either human or technological), the responder intervention phase, the decision phase, and the movement phase. A series of scenarios were developed that could start the whole process and they were categorized as man-made events, natural events, fire events, and security breaches. These events could be internal or external. The detection and process system could be separated into automatic or assessed. With automatic systems, an alarm goes off in response to whatever the event might be. With assessed systems, information is required so that someone can assess the nature or the existence of the incident and determine what the response should be. Generally, the group determined that as much information as possible should be gathered either by dedicated detection systems, CCTV, or human response to a command center in the building, and this information needs to be conveyed directly to the fire department or other first responders. This should provide the redundancy needed and should ensure that the people responding to the incident are well informed, which should reduce subsequent investigation time. This information could relate to the incident itself, the procedure, the population – where they're going, the level of impairment – and the structure itself. The group envisioned use of a CCTV system to monitor the movement of the people moving around in nonemergency situations. The group discussed the value of the notification system in both instigating and informing the response. The group decided that both visual and audible communication was preferred and would solve issues involving language or impairment. Guidance concerning the egress routes to use in the form of graphic displays over the doors indicating whether they are available and/or indicating the availability of routes and the direction of movement. This would require some standardization such as coloring and pictographs used. All systems would have to be adaptive and not hard-wired. The use of secondary devices such as PCs, TVs, cellphones, and PDAs also was discussed; these devices could be interrupted to ensure that people would be disengaged from their normal activities and received the message. The nature of the information passed to building occupants needs to be assertive and the time of the announcement needs to be clearly stated to emphasize that the event is real, not a drill. This announcement should be supported/reinforced by staff members. Considerable discussion was devoted to reducing the time it takes to get people to specific locations and floors. A consensus was not reached but it was determined that staff support will be needed. Accelerating the arrival of persons of authority on the critical floors was considered a top priority. It also was emphasized that structures should be people-centric – that is, designed around the occupants – and not just building a structure and then filling it with people. Information was at the forefront of all discussions as the most critical of issue in event response.

•   Group 2 - This group focused on the fire fighting process, and 11 process components were identified specific to high-rise buildings. The first one discussed was pre-planning for an event. Communicating these plans to the occupants and staff is vital since these are the individuals that will be on site prior to arrival of the responders. The event itself needs to be anticipated. The code process needs to be enforced to ensure that all systems are working, thus avoiding glitches during response. Detection, both human and technological, was discussed. Many alarm systems give good information (e.g., where the alarm is, what type of alarm it is, how many zones are involved, and what type of detector was activated) but the cause of the alarm nevertheless needs to be verified. This can be done with multi-detection systems. False alarms happen frequently and there have to be ways to deal with those (e.g., the assign of appropriate penalties for multiple false alarms). For this group, communication was also of critical importance. Timely communication providing adequate details is important from the moment that the monitoring company detects a problem. Gathering information from the staff and occupants also is crucial. Standards for the evacuation of

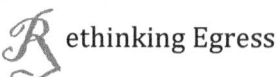

a building vary from place to place but a building plan is critical. Communication to the occupants in a manner that they cannot ignore is of tremendous importance. Some political questions also surfaced such as whether a smaller community with a small fire department and staff should be allowed to build tall buildings. The use of elevators for response and evacuation purposes also was discussed. The need for enhanced command center information was identified (e.g., to provide real time information from the incident site about such things as temperature on floor and smoke conditions). It was noted that hose deployment in stairwells is a continuing problem both in terms of blocking the stairwell and permitting contamination of the entire stairwell. Needed is some way to get a fire hose with positive pressure through a door while allowing the door to close so as to prevent stairwell contamination. With respect to ensuring that all building occupants have left, the group concluded that going door to door is still the only way to do it. It also was noted that post-incident information gathering, analysis, and archiving – or rather the lack thereof – by the fire department has been criticized but little has been done to improve the situation.

•   Group 3 - This group addressed "general" buildings where, at any given time, a large segment of the population would range in age from very young to very old and would be of widely diverse backgrounds and ethnicity. Retail malls were considered a good example of such occupancies. In terms of emergency event, the mall is more likely than most other occupancies to experience a wide variety of events (e.g., fire, criminal activity, terrorist activity, and severe weather). Communication was identified as key by this group and participants agreed that using cellphones would be a good way of notifying occupants of the event – perhaps by creating a special emergency ringtone used only for emergencies so it would be immediately recognizable. The designers of the malls and the retailers want you to see their stores; therefore, the exit signs are obscure. The visibility of exit signs needs to be enhanced as do the pathways to the exits. CBR detection should be incorporated into mall systems and severe weather should be announced over the PA system or using the cellphones. Elevators need to be on emergency power so that mobility impaired persons can easily exit. Risk also needs to be considered. There are different risks for different hazards requiring different solutions. In developing solutions, the risks that come with the various occupancies, heights, and other factors related to a specific structure should be considered. It was noted that the mall security staff probably would be the first ones to assess the situation and would be the ones to use for crowd control.

It was asked whether current cellphone technology permits a signal to be sent to all phones and it was explained that cell casting permits a signal to be sent to all phones within a single at the same time regardless of the number. Adaptations of this technology permit it to be directed only to a single building and some universities use it campus-wide. The FCC also is working on Cellular Mobile Alert Service (CMAS) standards which are now being evaluated by the industry for adoption and implementation.

•   Group 4 - This group focused on assembly buildings and the behavioral aspects of emergency evacuation. Assembly space was defined as a single large geographically complex area where occupants are generally aware of the egress routes because they are the same as those used for entry. Some discussion was devoted to comparing the impact of an event in an assembly space versus a high-rise building. People tend to receive their information in assembly space at the same time rather than at separate times which often happens in high-rises. Assembly spaces also tend to have large numbers of people in a single space and the need for well trained crowd managers is paramount. The group discussed this from an emergency planning aspect and a stadium was used

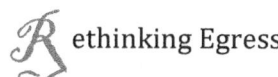

as an example. The consensus was that the most reasonable approach would be to divide the crowd into manageable subgroups. The highly trained crowd managers would be in constant contact with the emergency managers, usually via radio. The emergency managers then could determine what type emergency had occurred and direct the crowd managers to ensure that the subgroups were directed and shepherded to the right places. These managers could be the same persons that greet the people as they come in to the venue. Planning and training of the crowd managers obviously would be very important, but not much time should be needed after an event occurs to start moving the crows out to safety. Community emergency plans also were discussed in terms of their interface with building emergency plans. Ideally, if management fails, the architecture will serve to move people way from the event. The group also acknowledged that there are different strategies for seated versus non-seated venues. It was noted that very little attention has been paid to this type of occupancy in North America and that conferences on the topic would benefit from the participation of individuals like those attending the workshop

Once the participants convened again as a single group, Mr. Averill asked whether specific areas require long term research. The general consensus was that information and communication techniques are most in need of improvement. It was felt that this is where research, organization, and funding should be concentrated. Another point of view was that presenting technical aspects to a non-technical audience warrants more work. There are also many questions about cost, research could strive toward more cost-effective solutions and dual purpose solutions to satisfy clients and justify the cost. It was also noted that the technological ability to get information exists and this data needs to be displayed, in real time, in such a way that staff on the scene and then the first responders can evaluate the data. Another point brought up was that emergency egress is directly influenced by how the everyday people movement functions.

Mr. Averill pointed out that NIST has started a systematic program of data collection of fire drills in buildings. This covers multiple occupancies, heights, and counter flow issues. A matrix exists that we're trying to fill out over time in order to find out how these systems behave. Another aspect of the study is the human behavior issues. We have all the equipment to gather this data but have encountered problems gaining access to buildings. He continued that there are differences between fire drill data and real emergency data and of course right now only fire drill data is available.

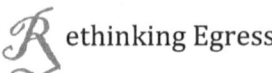

Mr. Averill reminded everyone that up until this point the discussion concentrated heavily on objectives and alternatives and the context of building egress. This data needs to be used in the process of moving forward, he stated.

The day was devoted to discussing the group's ideas of where the efforts and research agenda are best utilized. The lists of alternative suggestions were synthesized and many new innovative ideas came up. The list was divided into categorical ideas. Mr. Averill also suggested other topics for general discussion such as gaps, research areas, implementation, and research and development – in general things that may have been missed. Many topics were suggested generating a good cross section of research study material.

Richard Peacock of NIST explained how he and his staff took all 400+ alternatives and organized them into categories; means, methods, tactics, and strategies. The NIST staff organized the alternatives into "bins" and then placed the alternatives in the bins that most fit the various categories. Examples of bins were:

1. Sprinklers, areas of refuge, things related to design and initial construction of the building, solutions that would aid the fire service
2. Solutions external to the building such as sky bridges and external evacuation devices.
3. Flow of information in many directions as part of the evacuation process.
4. Making the existing system of egress more efficient.
5. Making the load of the system less in order to impact the evacuation system minimally.
6. Societal changes such as: avoid tall buildings, prevent law suits, transporters, etc.

The last exercise of the workshop was to separate into groups of 3-4 persons and go through all the ideas and rate them using a numeric scale from 1-9 on the viability of the various ideas. The participants were urged to draw on their expertise to rate the ideas and determine how different and "out of the box" the ideas are compared to current practices. The forms were passed out and everyone formed smaller groups. Questions were raised on the creativity issue and how needed that is. The response was that some of the more creative suggestions were not meant to be implemented immediately but to act as the foundation for serious research possibly not to be realized until decades from now.

The input was then averaged by NIST staff to determine how the group rated the alternatives. During this time there was continued discussion on the benefits of further research in all aspects of egress.

Mr. Averill mentioned that there will be a follow-up workshop that will be as inclusive as possible. There were many mentions of that this really is an international issue and there will be a lot of conferences in 2009 and 2010 dealing with related issues that this group could contribute to and benefit from.

Following the discussion Mr. Peacock summarized the evaluations showing the summary on the screen. Professor Keeney pointed out that the numerical responses were averaged based on the majority. On the feasibility scale a range was used. That does not summarize everything but many of the alternatives do have a range. Mr. Peacock continued that the results varied from the bottom of the scale to the top of the scale resulting in very divergent results. Those that were rated high can be considered "low hanging fruit" that the world should think about. Due to the scant time in

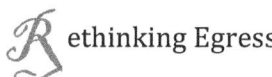

which the evaluation was conducted it was suggested that perhaps comments can be made after the meeting ends. It was also mentioned that many did not agree with the ideas as they were but they would agree with something close to it and that way some of the responses were generic. It was also suggested that the list could be modified into existing systems, feasible new systems, and social/economic issues. It was generally felt that the last form was vague.

The floor was turned over to Professor Keeney who expressed his delight in the breadth, depth of knowledge, and experience of the participants. He thanked everyone for their hard work and it appeared everyone had a genuine interest in creating the alternatives. This conference has resulted in a fairly comprehensive set of objectives for egress and a very broad range of alternatives he stated. This happens when everyone does individual thinking and then collective thinking. It's frequently difficult to articulate objectives and we provided a few techniques to assist and stimulate that. Creating alternatives is also tough and we came up with quite a few. He thanked everyone for their work and praised the NIST staff for their efforts.

Mr. Averill offered concluding remarks which summarized the workshop. He thought that the workshop had taken a very close and in-depth look at building egress and evacuation. The presentations were informative and enlightening. The workshop yielded a comprehensive set of 16 high-level objectives including the fundamental objectives and many alternatives. A few key concepts became persistent themes throughout the workshop, summarized below:

- Information is crucial. Gathering it. Processing it. Communicating it. In an era of change, no aspect of building evacuation is changing faster that information. With power, however, comes responsibility. Simply because we CAN collect a piece of information, does not mean that we should force feed that to our emergency responders, our building managers, or our occupants. It must be carefully triaged, assessed, and presented in a manner which incorporates what we know about human factors, social psychology, and situation awareness. Above all, however, information must not be ignored.

- Building evacuation procedures must be robust and adaptable. The procedures must adapt to (a) changes in the event and (b) different threats and scenarios.

- Design must make egress systems (first) obvious and intuitive and (second) integrated into the everyday use of the building, WITHOUT compromising our

FIGURE 2: 2008 RETHINKING EGRESS WORKSHOP PARTICIPANTS

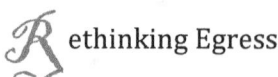

growing desire for physical security. Elevators may be an alternative which allows occupants in tall buildings to go out the way they came in, as well as provide egressibility to persons with mobility impairments, in the same way the ADA provided accessibility.

- Solutions which can apply to the existing building stock as well as our new construction will have greater impact in terms of the number of people and lives affected.
- Technology should be harnessed to deliver notification and induce direct action. Cell Casting, desktop computer interruption, PDA's, cable television, improved signage, and other modes should deliver reliable, timely, updated information targeted to specific people.
- We need to ensure accurate, timely, and reliable communications for firefighters. Firefighters need a way to communicate to the occupants and a way to obtain information from the occupants.
- As a community we need to perform cost-benefit analysis of many of the alternatives that exist. But that requires a measure of performance and an assessment of cost. In many cases we fall short of both.
- Finally, evacuation will not always be the preferred solution. Protect in place, phased evacuation, and areas of refuge may minimize business interruption, as well as minimize harm to occupants. The nature of the threat will dictate the appropriate response.

Mr. Averill then thanked the participants and NIBS staff and declared the 2008 Rethinking Egress workshop concluded.

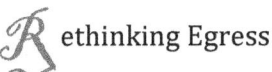

WORKSHOP FINDINGS

The findings of the workshop are presented here in the form of the compilations of emergency egress objectives and alternatives identified.

CONSOLIDATED EMERGENCY EGRESS OBJECTIVES

Save Lives
      Maximize survivors
      Staying alive
      Mitigate consequences of events (death, injury, stress)
   Save occupant lives
      Save lives of occupants
   Save responder lives
      Save lives of first responders
      Minimize risk to life of external agencies required for evacuation
      Preserve emergency responder life safety

Prevent Injury
   Prevent occupant injury
      Minimize the harm to all people during evacuation
      Minimize the injury level (short term) of survivors in exposed population
      Minimize the injury level (long term) of survivors in exposed population
   Prevent responder injury

Minimize or reduce risk of injury (short term) of external agencies

Minimize or reduce risk of injury (long term) of external agencies

Minimize Economic Costs
   Reduce societal cost of loss of lives
   Reduce societal cost of protecting lives
   Reduce business cost of protecting lives
   Ensure cost effective construction and operational costs
   Reasonable costs – first and continuing
   Minimize or reduce the cost in terms of disruptions caused by evacuation

Minimize or reduce the cost of enabling the actual evacuation

Use no more resources than society is willing to expend.

Provide solutions that are cost effective

Decrease cost of building and variety of buildings.

Reduce egress costs and carbon footprint.

Maximize return.

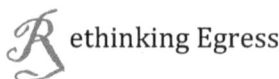

Decrease cost of the building.
    Prevent law suits.

Minimize Property Damage
    Improve first responders' abilities to safeguard property
    Save properties
    Preserve property.

Minimize Impact on Property Operations

Prevent loss of business.

Minimize impact on building operations
    Minimize negative impact on the functionality of the building
    Allow normal functioning of building.
    Avoid unnecessary disruptive evacuations.

Reduce Stress
    Minimize post traumatic stress
    Reduce stress on occupants
    Reduce stress on emergency staff
    Minimize or reduce physical and mental stress on evacuees
    Reduce panic
    Move people without undue stress.
    Minimize possibility of panic and other destructive behaviors
    Prevent confusion.

Reduce Grief to Relatives and Friends

Enhance Communication
    Increase confidence in message and communication system
    Develop trust in the message and communication system
    Allow responders to communicate effectively within building.
    Notify all occupants in structure of emergency and need to evacuate.
    Improve communication to building occupants.
    Provide the most detailed information possible to occupants.

Minimize Response Time
    Assure appropriate response

Minimize Evacuation Time
    Reduce time
    Minimize total egress time
    Decrease evacuation time
    Provide high-capacity/short time egress
    Enable immediate start of evacuation
    Simplify egress process
    Speed up building evacuation
    Provide efficient egress system

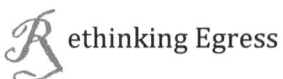

Create efficient handicap egress
Provide equality for egress of disabled and able bodied occupants.
Develop methods to increase speed of evacuation.
Quickly move/guide population away from harm (smoke, fire, bio, chem.).
Ability to coordinate an orderly exit
Provide uncompromised pathways
Provide the earliest possible notification.

Enhance Safety
Increase safety, irrespective of location
Make evacuation process safe
Keep occupants safe
Provide safe egress for all occupants
Provide safety for all occupants regardless of age to mobility

Maximize life safety.
Increase safety of community.
Provide safety for all occupants.

Provide safety for most occupants.
Provide safety from fire for all occupants.
Ensure that on one is left behind.

Ensure timely rescue of injured or 'at risk' occupants
Ability to move vast amounts of people safely

Move people to a safe place for a range of threats

Reduce fire department interference with occupants evacuating building

Move people to a safe place with a range of people

Remove people from imminent danger.
Enhance responder safety.
Retain flexibility to rescue other building occupants if conditions change
Insure system works for different types of emergencies/threats.
Allow escape without aid of fire fighters.
Operate during failure of city utilities (water or electricity).
Ensure it can be maintained

Make People Feel Safe
People feel safer
Enhances "sense of safety"
Provide a Safer Environment
Give occupants a sense of security.
Provide occupant confidence.
Increase confidence in message and communication system

Maximize Situation Awareness

Provide real-time information to incident command and to occupants
Assess available information to enable management of occupants and staff
Incorporate knowledge of the specific event.
Provide real-time information to incident command and to occupants

Facilitate Responder Access
    Enable responder access to problem area(s)
    Provide access to fire fighting operations
    Facilitate fire-fighting
    Separate fire-fighting from evacuation

Clear building area for fire fighting.
    Provide for safe and effective fire fighting operations.
    Free up responders
    Maintain uncompromised pathways.
    Allow the FD to do their job – control the fire, prevent collapse, etc.

Improve Education and Training
    Manage behavior of population to ensure safety, comfort, and security
    Increase population's perception of risk
    Enhance timely occupant action in an emergency

Improve Knowledge about System
    Increase knowledge of the occupants.
    Understand requirements/capabilities of population not to make situation worse, put them in more danger,
    Develop a framework to evaluate solutions on a risk basis for various scenarios.
    Provide verifiable and quantifiable benefit.
    Provide solutions that stakeholders accept.

Isolate Fire
    Improve fire service use of technological systems.
    Improve fire fighting to reduce egress needs.

Enhance Detection

Allow for quicker detection of the fire.

Provide Easy Egress Procedures
    Provide a resilient and simple system.

Reduce complexity and increase reliability
    Obviate the need for special skills and special equipment on part of occupants
    Ensure egress system meet k.i.s.s. principal.
    Simplify procedures
    Ask occupants to do only things they understand
    Ask occupants to do only things with which they are familiar
    Reduce space set aside for emergency use alone
    Increase dual or multi functionality.

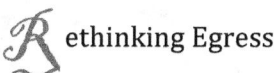

Flow with normal or expected human behaviour

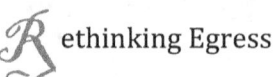

CONSOLIDATED ALTERNATIVES TO IMPROVE EMERGENCY EGRESS

ALTERNATIVES INVOLVING SPRINKLERS / ACTIVE SUPPRESSION SYSTEMS

Use sprinklers and compartmentation
Redundant water supplies
Mandate sprinklers in all building, except residential
Retrofit sprinklers in all high-rise buildings
Increase sprinkler spray design basis to suppress a full-floor fire
Increase sprinkler density with building height
Improve reliability of existing building water supplies and sprinklers
Less trade-offs for sprinkler systems
Increase sprinkler reliability
Improve sprinkler system design
Mandate sprinklers for all buildings – including residential

ALTERNATIVES INVOLVING PROTECT IN PLACE / AREAS OF REFUGE

Increase number of places of safety
Areas of rescue assistance in stairs
Transitional refuge areas for resting and transition to elevators in super tall buildings
Safety zones
Protected queuing spaces
Zoned floors to create "safe" areas for people awaiting evacuation
Areas of safety
Providing areas with safe sections
For very large, or very tall, or very complex buildings → fully protected, adequately lit, properly ventilated refuge spaces of adequate size and a means of 2-way communication with the emergency response team
Protect in place (no egress)
Provide alternative areas for egress assistance.

ALTERNATIVES INVOLVING BUILDING CONSTRUCTION CHANGES

Lightweight innovative materials for egress paths
Control fuel load to avoid evacuation
Enhanced fire resistive construction for high-rise buildings
Reduce fuel loads in buildings
Limit use of combustible materials
Self-protected buildings – sprinklers, fire/smoke resistant construction
Prevent fires from occurring in the first place through material selection and control
Build Buildings to Egress Standards

ALTERNATIVES INVOLVING BUILDING MATERIAL CHANGES

Protected corridors, stairways, skybridges
Harden existing buildings
Develop new structure external to existing buildings
Place exit signs high and low
Battery backup for egress lighting
Hardened evacuation routes
Design buildings not to collapse
Shorten evacuation routes
Mechanize evacuation routes (moving walkways, elevators)
Improve quality of egress routes – finishes, natural light, view
Adequate lighting
Adequate width/capacity of egress
Adequate signage
Smoke control system with training for fire department use
Protected exit route
Install readily visible directional indicators at key locations
Normal and emergency movement systems should be integrated
Egress systems designs that reflect human behavior
Pressurized stairwell
Effective smoke control
Move standpipes
Separate emergency cores
Create horizontal and vertical compartmentation
Use horizontal dimension for egress rather than rely on vertical
Improve visibility of signage
Require battery back-up and emergency generators
Increase use of daylight through windows and skylights
Build-in ergonomics to system designs
Require separation of egress paths
Make exits attractive, functional and normally used
Color stairs make them more user friendly
Alternate discharge at base of building
Horizontal exits
Building subdivision
Confining smoke, heat, and fire to a small portion of the building
Fire/smoke isolation of elevators the lobbies
Flashing exit signs
Require/provide manual and automated vent opening in buildings
Provide hardened floors at intervals in tall buildings
Durable (temp, smoke, impact) enclosure envelope
Enhance illumination
Ensure smoke control has an emergency power supply or that all dampers are fail safe
Improve signage
Provide large "fire doors" to compartmentalize large floor plans
Prevent smoke from entering stairs
Structure the building for dispersion (allow for evacuation without competition)
Enhance visibility of exit signs ▫ larger, flashing or digital signage
Enhance paths to exits with pavement markings

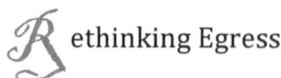

Locate exits in generally used areas such as toilets
Place painted or tiled lines to the exits
Flashing or re-designed signs

ALTERNATIVES INVOLVING THE FIRE SERVICE

Access for fire fighters independent of occupant egress
Simple to use protective breathing apparatus
More effective emergency responders
Do not bring fire fighters into tall buildings
Equipment lift for fire fighters
Fire fighter elevators
Robotic fire fighting equipment
Wireless technology to monitor fire command station en route to incident
Provide building repeaters systems for portable radio coverage with tested and maintained system
Dedicated fire fighter ingress
Get firefighters out of stairs
Rapid fire service intervention
Responder's stairwell
Prepositioned assets
Separate firefighter access from occupant egress
Increase training and education for firefighters
Provide on-site fire brigades for high-occupancy buildings and complexes
In-house fire brigade will obviate need for outside firefighters
Increase funding for R&D on firefighter equipment / technology
Dedicated firefighting stairwell
Better ventilation of to support firefighting
Adequate staffing and equipment to handle emergency
Building conditions available at command center
Wireless communication that can be relayed from building to command post
Train firefighters on new technological systems
Designated assembly areas for emergency responders
Mandate onsite training of firefighters
Provide reference materials in fire houses
Real-time building / fire data to responding fire service personnel
Standardized, flexible system to communicate with occupants and emergency responders using
conventional communication devices and technology (COMM)
Capability to maintain incident command from remote location
Fighting fires from adjacent buildings
Training of responders for occupant response (PLAN)
Train fire service to operate building fire response systems (PLAN)
Have a local fire company for large buildings
Provide fire fighting vestibule at each story – outside of the stairs
Provide air to recharge oxygen bottles for fire fighters in high-rises
Install a dumb waiter in stairway or elevator shaft for FF equipment (not personnel) and Install a
smaller (30") stairway for fire fighter access
Position fire brigades within high-rise buildings (with occupancies greater than 10,000) that are
trained for managing egress as well as fighting fires. These could be private or city employees.

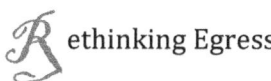

Additional skills, with value to the building owner or occupants, would be exploited for times when no emergency was present
Vestibules for standpipe to fight fires
Improve fire department training
Provide mechanical means to allow fire hose to penetrate stairwell doors while keep smoke from entering
Cell phone notification to emergency responders

ALTERNATIVES INVOLVING ELEVATORS

Simultaneous use of elevators and stairs
Use elevator or other mechanical means
Protected elevators for disabled
Protected elevators for evacuation of transitional refuge areas
Hardened elevators
Protected elevator lobbies
More aggressively educate officials, owners, and providers about the benefits of protected elevators
Use elevators
Provide incentives to increase elevator use
State of the art elevators
Training and drills in the elevators (maybe video training)
Use of elevators for egress
Require elevators vestibules in all buildings with elevators
Provide equipment hoist/shaft for fire fighting equipment
"Hardened" elevators with reliable, alternative power
Improve elevator doors to reduce smoke stratification in buildings – may increase evacuation time
Make use of elevators accepted by the public
Lock out and tag out elevator power

ALTERNATIVES INVOLVING SOCIETAL /REGULATORY / LEGAL CHANGES

Don't build high buildings
Develop risk "scale" for building occupants
Bill for response based on risk
Incorporate response costs in investment, design and development cost
Revalue sprinklers (in real costs)
National standards for crowd management and evacuation training
Tear down tall buildings
National high-rise building codes that require modern egress
Increase integration of rules such as emergency and security to reduce conflicts
Quantitative performance criteria
National standard for emergency action protocols
Make occupants more fit
Staggered density in buildings
Provide immunity from lawsuits for research-driven approaches
Standardize evacuation procedures across country by emergency type
Prevent lawsuits

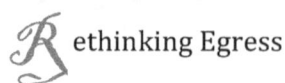

Standardize inspection program and testing
Limit tradeoffs
False alarm requirements with penalty for false alarms
Legislate immunity for performance-based design
Require regular testing and inspection
Required modeled systems to be tested after installed
Look at evacuation from an urban scale
Develop threat strategies and evacuation strategies
Share emergency systems with multiple buildings
Standardize egress strategies according to type of emergency
Annual tests of emergency systems: generators under full load, fire pumps, egress
Restrict the heights of tall buildings
More research money for egress studies to obtain more realistic and up to date data
Analyze time necessary to evacuate building safely prior to spending money
Improve public education on eating healthy to reduce obesity problem in this country
Limit assembly occupancies at the top of buildings
Require periodic tests of smoke control systems
Don't reduce the fire-resistance of building elements when sprinklers are installed
Ensure decision makers consider cost/benefit analysis for each alternative
Change standards and regulations

ALTERNATIVES EXTERNAL TO THE BUILDING

Chutes / slides / controlled descent devices
Horizontal egress or skybridge
Transporters
External evacuation devices
Use platforms and chutes for external evacuation
Provide skybridges to connect buildings together
Skybridges every 5 floors – many buildings linked together
External devices
Horizontal / skybridges
Use bridges between buildings to create horizontal exits
Explore use of external evacuation systems
Bridges to other buildings
Connections – permanent or temporary – to adjacent buildings
Create egress 'lifeboats' or escape pods
Provide helicopter sites at all tall buildings
Install deployable bridges in new buildings that can couple, with minor modifications, to adjacent existing buildings
External devices, platform devices
Personal egress modules for external egress

ALTERNATIVES INVOLVING COMMUNICATION

Highly motivating alarm/alert/management systems
Communications to and from all parties

Standard visual emergency communication system
Emergency notification system to all occupants – phone, email, TV, satellite, public address
Egress communication system
Utilize all methods of communications in an integrated network
Earlier notification of the need to move
Simple, effective communication
Robust communications
Active egress management systems
Communicate instructions to occupants to steer the evacuation
Design interoperable emergency communication systems amongst responders
Design real-time communication to maximize effectiveness
Use technology such as cell phones, PDA's, computers, phones for timely communication
Two-way communication between fire service and occupants during event
Two-way communication in stairwell
Include security for fire alarm devices
Broadcast instructions over cell phones
Video broadcast to cellular with voice over instructions
Use loud and intrusive communication modes (video and audio)
Real time pictorial emergency communication system
Multiple communication messaging systems
Two-way communication between fire fighters / responders and occupants
Protect fire notification system
State of the art communication systems
Situation awareness
Symbolic / iconic notification system for incident type and action required of occupants
Improve pictorial representation of emergency communication (international symbols)
Integration of communication systems into existing buildings
Communication – 2-way between occupants and first responders
Broadcast alarms/warnings on cell phone networks at building
Broadcast alarms to computer networks in building
Initiate alarm/message earlier to people – so that elevators could be used for longer
Notify occupants via cell phone
Universal emergency ringtone
Use sound via cell phone to guide to exits
Language preference for cell phone/PA announcements
Regular announcements that include emergency egress information
Announcements in multiple languages

ALTERNATIVES INVOLVING INFORMATION SYSTEMS

Building guidance systems
Distribute relevant information in real-time
Real-time signage used on daily basis that results in high level of occupant acceptance
Enhanced information to building occupants and management
Intelligent alerting and response decision-making software
Print emergency egress instructions on the back of notepads, placemats, other locations that are readily available to transient occupants.
Provide multi-lingual instructions

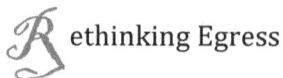

Have pre-recorded messages ready for use.
Utilize dynamic signage to provide information
Dynamic reactive signage
Precise information
Innovative means to provide information to people (cell phones, blackberries)
Real-time reporting of conditions
Visualization technology
Facilitate milling
Better situational information
Use public service announcements to increase perception of risk
Increase usage of icons, pictographs and signage to direct occupants
Real time monitoring of movement in stairs
Occupancy badge and readers so you know where people are
Provide remote monitoring and control of large building fire systems
Provide people with the right messages (content, style, etc)
Real-time information to occupants about what is going on in the building (where the smoke is, where others are, where FF are)
Personal monitoring for people with disabilities
Increase the elements required to be monitored
Information should be transmitted to the responding group (fire department, etc). It should refer to the incident (nature, existence), procedural status, population activities, structure (exit use, stair, route loss, etc.)
Multi-modes – visual, aural; should make use of traditional notification, visual systems, PC screens, TVs, PDAs, cell phones, etc.
Graphical screens could be used to maximize the information that can be provided.
Critical that information is consistent, dynamic, comprehensive and adaptive.
Live voice, provided by someone of authority

ALTERNATIVES INVOLVING PRE-EVENT PLANNING

Common plan of action for everyone: everyone evacuates by stairs
More effective emergency staff
Clear evacuation plans
Consider normal use of building
Train staff
Practice egress frequently
Interact with evacuees
Focus on one plan
Educate and train occupants
Emergency planning with programmed response
Make it clear there are other hazards than fire
Pre tested evacuation routes
Occupant fire safety training
Engage occupants during training to improve procedures
Create a mechanism of learning about the building environment and systems that is fun
Active management of egress process
Make stairs part of the normal building use
Web-based training

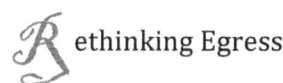

Require egress drills
Education
Educate and train building emergency managers
Mandate evacuation drills for multiple scenarios
Communicate preplan to fire fighters and occupants
Have preplan available during response
Reevaluate preplan on periodic basis
Create training courses for fire service and stakeholders
Involve occupants in solutions for egress from buildings
Training of building occupants related to building communication systems
Mandatory fire drills
Require fire drills for building occupants more often
Require more realistic fire drills – actual movement and alarms
Require fire drills for spaces with casual users (malls, retail, bars/restaurants)
Concise, meaningful pre-plans not only on file but pre-reviewed by potential incident responders
Provide incentives for occupants to participate in evacuation drills, e.g., a trophy to the work unit that had the most employers climb and/or descend the most number of stairs. This could be tied to a physical fitness program
Train/educate/practice all procedures
Train/educate occupants NOT to delay and tell them why
Present occupants with actual simulation/videos of what could happen in a real event
Have occupants go through FF training – experience a real event
Participate development of plans and procedures/education/training → involve them in the process
Conduct pilot project to demonstrate
Exposing occupants (in a safe way) to actual fire conditions (e.g., theatrical smoke)
Floor wardens must be people with some authority.
Subgroup managers should be incorporated into plans for assembly buildings – to lead the people out of the building
Heavy training of subgroup managers
Subgroup managers need to stay in constant communication with the emergency manager
Causing egress via a noxious odor or intolerable sound
Develop alternative egress plans

ALTERNATIVES INVOLVING EFFICIENT USE OF EGRESS SYSTEM

Enhanced stairway descent devices
Evacuation aids for disabilities
Photoluminescent markings
Building wayfinding/guidance system
Overhead rails for disabled movement systems in stairs
Set door swing to stairs to not interrupt flow
Disable email systems for more rapid initiation of egress
Gravity-driven, lightweight vehicles for mobility-impaired individuals
Create multi-use emergency routes
Equalize loads on all evacuation routes
Motorized egress devices for people with disabilities

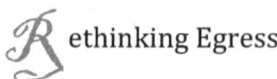

ALTERNATIVES TO ENHANCE STAIRWELL EVACUATION / REDUCE STAIRWELL LOAD

Multiple and wide staircases
Install tracks in stairwell handrails for trolley seats to be used by all occupants
Use electroluminescent strips in hotels
Recalculate stair widths
Set doors to stairs to allow for traffic flow
Positive pressure for stairwells
Increase stair capacity
Increase number of stairs
Force stair usage in everyday activities
Use vestibule for hose line deployment without contamination of stairwell
Paint stairs in white paint
Widen the stairs.
Remove fire equipment from stair towers
Design stairs with fire fighting procedures in mind
Positive air pressure in stairs
Wider stairs
Dedicated egress stairs for responders
Develop periodic 'rest areas' within stair towers
Require some level of hardening at exit (stair) enclosure
3rd stairwell for fire fighters

ALTERNATIVES IMPACTING THE DESIGN PROCESS / RISK INFORMED PERFORMANCE-BASED DESIGN

Base designs on risk assessments
Model-based evaluation of designs
Standardize design basis events and performance objectives to set public policy
Develop cost/benefit tools to deal with rare, high consequence events
Use modeling to estimate evacuation time for buildings for different circumstances
Consider worst-case scenarios and build in countermeasures
Use modeling to estimate evacuation time for buildings for different circumstances
Use egress modeling to choose between alternative evacuation strategies in response to designated scenarios
Perform engineering analysis to evaluate emergency procedures
Improve accuracy of evacuation software
Egress models should consider more threats than just fire
Perform cost/benefit analysis of solutions
Publish articles on scenarios which need to be considered
Design for hazmat and WMD, including chemical detectors
Use performance-based criteria for all threats, more than just fires
Develop process for creating threat scenarios and performance basis for providing alternatives acceptable to stakeholders
Base design on risk and performance-based design
Develop an understanding of all of the events that would prompt full building evacuation
Establish risk factors

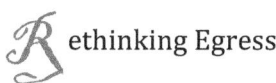

ALTERNATIVES THAT ENHANCE RELIABILITY OF BUILDING SYSTEMS

Leave power on for egress
Low exit signs
Regular inspections
Coordinated systems control
Better lighting
Sensing systems should be fail-safe
Provide redundant systems
Enhance reliability of sprinklers by electronic monitoring of flow
Design detection systems to reject false alarms
Constantly use systems to ensure it will work during an emergency
Increase number of escape routes
Backup power for evacuation components
Electronic monitoring of all components of standpipe
Two remote sources for sprinkler water
Monitor all life safety systems
Maintenance of integrity of egress paths
Redundant passive as well as automatic protection
Meaningful quality control before opening of the facility
Meaningful follow-up quality assurance throughout the life of the building at reasonable frequency
Document, calculate, observe construction and fully accept ultimate responsibility
Provide redundant water and electrical supply independent of city utilities
Increase monitoring of building safety systems.

ALTERNATIVES THAT INVOLVE EVENT PROCEDURAL CHANGES

Believable defend in place strategy
Ensure egress is only in response to real incident
Ensure equal use of egress routes
Design emergency procedures effectively
Establish evacuation procedures for disabled occupants
Guidelines for occupant evacuation
Stairs to refuge floors, elevators from there
Emergency procedures manuals
Decision support systems
Design unique procedures for each specific type of emergency
Allow use of "normal" elevators as long as possible
Standardize evacuation strategies
The use of staff in a more significant role in guiding occupants out of the building
Better use of fire wardens – maybe they are the first ones out (follow the leader)
Occupant procedures that actually take into account what people WILL do
Practice egress with real event scenarios – blocked stairs, theatrical smoke, FD in stairs
If we know that milling will occur, establish groups (evacuation groups of the occupants) to facilitate the process (since we know that they will likely get into groups anyway)

ALTERNATIVES THAT ENHANCE EVENT DETECTION

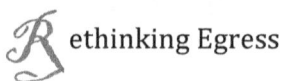

Timely notification from detector to monitoring company
Include adequate details on type and location of the event / detector
Use thermal imaging systems for detection
Automatic detection should be implemented for the spaces outside a main assembly area
Detect CBR
Detect severe weather

OTHER ALTERNATIVES

Locator badges with counters at stairs
Improve understanding of where people are in the building
Motivation of people to evacuate
Systems to motivate action
Use egress systems on a regular basis
More egress routes
More building exits
Enhanced notification systems at the local level
Require at least one remote path of vertical egress when a central core is utilized
Use devices to assist in compartmentalization
Provide individual protective gear for occupants

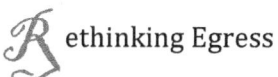

INDIVIDUAL EVALUATIONS OF ALTERNATIVES

The alternatives were evaluated on three distinct criteria:

- Usefulness: How good, in terms of achieving the fundamental objectives of egress, is the alternative IF it could be effectively implemented? Use a 1 to 9 evaluation scale where 1 is poor and 9 is great.

- Feasibility: What is the chance that the alternative could be effectively implemented within 10 years? Use a percentage evaluation scale, where 0 means there is no chance, 50 means there is a half chance, and 100 means that the alternative could surely be implemented.

- Creativity: How creative or innovative is the alternative? Use a 1 to 9 evaluation scale where 1 is a standard well-known idea and 9 is an idea that you had not heard of before.

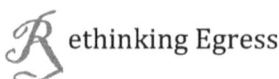

Useful (1-9)	Feasible (0-100)	Creative (1-9)	Alternatives
9	20,100	2	Retrofit sprinklers in all high-rise buildings
7	35,99	3	Transitional refuge areas for resting and or transition to elevators in super tall buildings
1,7	5,80	3,9	Provide individual protective gear for occupants
7	50,100	3	Hardened evacuation routes
1,7	5,95	5	Mechanize evacuation routes (moving walkways, elevators)
8	75,100	2,9	Install readily visible directional indicators at key locations
8	25,100	7	Normal and emergency movement systems should be integrated
5,8	50,80	4	Build-in ergonomics to system designs
7	30,100	6	Alternate discharge at base of building
9	40,100	3,7	Provide hardened floors at intervals in tall buildings
8	40,100	3,7	Enhance visibility of exit signs → larger, flashing or digital signage
2,8	0,90	2,8	Locate exits in generally used areas such as toilets
1	0,10	1,9	Do not bring fire fighters into tall buildings
9	60,100	4,8	Real-time building / fire data to responding fire service personnel
1,7	5,90	2,7	Install a dumb waiter in stairway or elevator shaft for FF equipment (not personnel) and Install a smaller (30") stairway for fire fighter access
5,8	20,85	7	Position fire brigades within high-rise buildings (with occupancies greater than 10,000) that are trained for managing egress as well as fighting fires. These could be private or city employees. Additional skills, with value to the building owner or occupants, would be exploited for times when no emergency was present
9	30,100	7	Provide mechanical means to allow fire hose to penetrate stairwell doors while keep smoke from entering
9	60,90	5	Simultaneous use of elevators and stairs

1,8	0,30	2,8	Restrict the heights of tall buildings
2,7	0,60	3,8	Share emergency systems with multiple buildings
3,7	0,80	2,8	Limit assembly occupancies at the top of buildings
2,8	5,85	7	Skybridges every 5 floors – many buildings linked together
3,8	20,85	7	Install external evacuation devices, e.g., platform devices
9	75,100	7	Emergency notification system to all occupants – phone, email, TV, satellite, public address
8	70,100	6	Two-way communication between fire service and occupants during event
7	40,100	4,7	Symbolic / iconic notification system for incident type and action required of occupants
9	50,100	8	Notify occupants via cell phone
3,8	80,100	9	Use sound via cell phone to guide to exits
9	60,100	5,8	Real time monitoring of movement in stairs
8	80,100	3	Photoluminescent markings
2,7	0,80	6,9	Install tracks in stairwell handrails for trolley seats to be used by all occupants
8	5,100	4,8	Base designs on risk assessments
7	5,90	5	Standardize design basis events and performance objectives to set public policy
7	5,100	3,7	Use performance-based criteria for all threats, more than just fires
8	70,100	3	Electronic monitoring of all components of standpipe
7	10,100	2,7	Ensure equal use of egress routes
3,8	0,85	3,8	Practice egress with real event scenarios – blocked stairs, theatrical smoke, FD in stairs

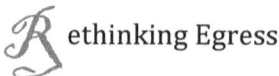

ACKNOWLEDGMENTS

The Rethinking Egress workshop was the combined effort of a significant number of people, including:

- Shyam Sunder, NIST
- Mat Heyman, NIST
- William Grosshandler, NIST
- Richard Bukowski, NIST
- Erica Kuligowski, NIST
- Barbara Faverty, NIST
- Claret Heider, NIBS
- Carita Tanner, NIBS
- Ralph Keeney, Duke University
- Najib Abboud, Weidlinger Associates
- Carl Galioto, Skidmore, Owings, and Merrill
- Clas Jacobsen, United Technologies Research Center
- Dennis Mileti, University of Colorado – Boulder
- Russ Sanders, National Fire Protection Association
- Harold "Bud" Nelson (ret.)
- The staff at the Airlie House Conference Center
- Brad Pumphrey, EastBay Media

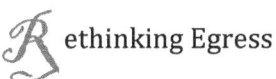

APPENDIX A: LIST OF ATTENDEES

Najib Abboud
   Weidlinger Associates, Inc.

Carl Baldassarra
   Schirmer Engineering

Richard Bukowski
   NIST

Ron Cotè
   National Fire Protection Association

Michael Dillon
   Dillon Consulting Engineers, Inc

David Frable
   U.S. General Services Administration

Carl Galioto
   Skidmore, Owings, and Merrill LLP

Steve Gwynne
   Hughes Associates, Inc.

Claret Heider
   National Institute of Building Sciences

Clas Jacobsen
   United Technologies

Peter Johnson
   Arup

Ralph Keeney
   Duke University

Michael Lindell
   Texas A&M University

Dennis Mileti
   University of Colorado, Boulder

Jake Pauls
   Jake Pauls Consulting Services in Building Use and Safety

Jason Averill
   NIST

Richard Bowers
   Montgomery County Fire and Rescue

Diane Copeland
   Dillon Consulting Engineers, Inc.

Geoff Craighead
   Securitas Security Services USA, Inc.

Thomas Fitzpatrick
   Guiliani Partners,LLC

Edwin Galea
   University of Greenwich

William Grosshandler
   NIST

Glenn Hedman
   University of Illinois at Chicago

Matthew Heyman
   NIST

Keith Johnson
   Fairfax County Fire and Rescue

Edwina Juillet
   Consultant

Erica Kuligowski
   NIST

Nancy McNabb
   National Fire Protection Association

Harold "Bud" Nelson
   Formerly of NIST (Retired)

Richard Peacock
   NIST

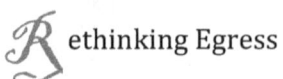

Guylene Proulx
National Research Council – Canada

Jonathan Shimshoni
Escape Rescue Systems, Ltd.

Carita Tanner
National Institute of Building Sciences

Steven Winkel
The Preview Group, Inc.

Jerry Woolridge
Reedy Creek Improvement District

Russell Sanders
National Fire Protection Association

Janne Sorsa
Kone Elevators, Ltd.

Keith Wen
New York City Department of Buildings

Antony Wood
Council on Tall Buildings and Urban Habitat

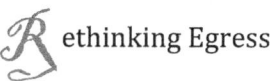

RETHINKING EGRESS WORKSHOP
Airlie House Conference Center, 809 Airlie Road,
Warrenton, Virginia 20187

April 1 – 3, 2008

This workshop has been convened by the National Institute of Standards and Technology (NIST). It was organized by the Multihazard Mitigation Council (MMC) of the National Institute of Building Science (NIBS) with Funding from NIST.

Tuesday, April 1	
8 to 9 am	Breakfast
9 to 10 am	Welcome and Overview
10 to 11 am	Identification of Current Alternatives
11:30 am to 1 pm	Lunch Break (lunch served from noon to 1 pm)
1:30 to 5 pm	Guided Brainstorming
5 to 6 pm	Free Time
6 to 7 pm	Pre-dinner Social Hour (cash bar)
7 to 8 pm	Dinner
Wednesday, April 2	
8 to 9 am	Breakfast
9 to 11:30 am	Alternatives from Objectives
11:30 am to 1:30 pm	Lunch Break (lunch served from noon to 1 pm)
1:30 to 5 pm	The Process of Evacuation: More Alternatives
5 to 7 pm	Free Time
7 pm	Dinner
Thursday, April 3	
8 to 9 am	Breakfast
9 to 11:30 am	Identification of Gaps and Workshop Summary
12 noon to 1 pm	Lunch served
Transportation to Dulles Airport will depart from Airlie House at approximately 1:30 pm.	

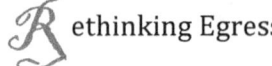

Name: _____

1

List All Ideas that You Want to Have Considered at the Workshop

A. _____

B. _____

C. _____

D. _____

E. _____

F. _____

G. _____

H. _____

I. _____

J. _____

K. _____

L. _____

M. _____

N. _____

O. _____

P. _____

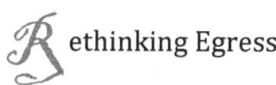

2

Name: _____

List All of the Alternatives that You Think May Enhance Egress

A. _____

B. _____

C. _____

D. _____

E. _____

F. _____

G. _____

H. _____

I. _____

J. _____

K. _____

L. _____

M. _____

N. _____

O. _____

P. _____

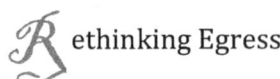

3

Name: _____

List All of the Objectives that You Think May Enhance Egress

A. _____

B. _____

C. _____

D. _____

E. _____

F. _____

G. _____

H. _____

I. _____

J. _____

K. _____

L. _____

M. _____

N. _____

O. _____

P. _____

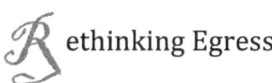

Name: _____

Create a List of All Values that You Care about Concerning Egress

A. _____

B. _____

C. _____

D. _____

E. _____

F. _____

G. _____

H. _____

I. _____

J. _____

K. _____

L. _____

M. _____

N. _____

O. _____

P. _____

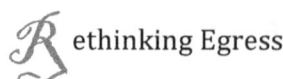

Name: _____

Expand Your List of Values that You Care about Concerning Egress

A. _____

B. _____

C. _____

D. _____

E. _____

F. _____

G. _____

H. _____

I. _____

J. _____

K. _____

L. _____

M. _____

N. _____

O. _____

P. _____

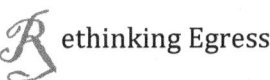

6

Name: _____

Convert Each of Your Values (On Forms 3 and 4) to Objectives (Verb and Noun)

<u>Basis</u>   <u>Objective</u>

A. _____   _____

B. _____   _____

C. _____   _____

D. _____   _____

E. _____   _____

F. _____   _____

G. _____   _____

H.

I. _____   _____

J. _____   _____

K. _____   _____

L. _____   _____

M. _____   _____

N. _____   _____

O. _____   _____

P. _____   _____

Name: _____

Create Alternatives, or Elements of Alternatives, Using Each Objective

<u>Basis</u>   <u>Alternative</u>

A. ____ _____

B. ____ _____

C. ____ _____

D. ____ _____

E. ____ _____

F. ____ _____

G. ____ _____

H. ____ _____

I. ____ _____

J. ____ _____

K. ____ _____

L. ____ _____

M. ____ _____

N. ____ _____

O. ____ _____

P. ____ _____

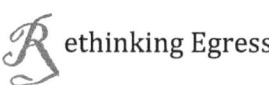

Name: _____

Create Alternatives, or Elements of Alternatives, Using Each Collective Objective

<u>Basis</u>   <u>Alternative</u>

A. _____   _____

B. _____   _____

C. _____   _____

D. _____   _____

E. _____   _____

F. _____   _____

G. _____   _____

H. _____   _____

I. _____   _____

J. _____   _____

K. _____   _____

L. _____   _____

M. _____   _____

N. _____   _____

O. _____   _____

P. _____   _____

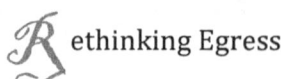

Group Names: _____

9

Create Alternatives, or Elements of Alternatives, Using Each Collective Objective

A. _____

B. _____

C. _____

D. _____

E. _____

F. _____

G. _____

H. _____

I. _____

J. _____

K. _____

L. _____

M. _____

N. _____

O. _____

P. _____

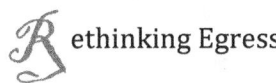

Name: _____

List All Components of an Effective Egress Process

A. _____

B. _____

C. _____

D. _____

E. _____

F. _____

G. _____

H. _____

I. _____

J. _____

K. _____

L. _____

M. _____

N. _____

O. _____

P. _____

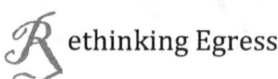

Name: _____

List All Components of an Effective Firefighting Process

A. _____

B. _____

C. _____

D. _____

E. _____

F. _____

G. _____

H. _____

I. _____

J. _____

K. _____

L. _____

M. _____

N. _____

O. _____

P. _____

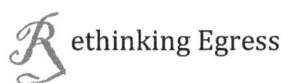

Group Names: _____ **12**

List Any Additional Components of an Effective Egress Process

A. _____

B. _____

C. _____

D. _____

E. _____

F. _____

G. _____

H. _____

I. _____

J. _____

K. _____

L. _____

M. _____

N. _____

O. _____

P. _____

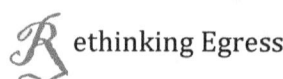

Group Names: _____ **13**

List Any Additional Components of an Effective Firefighting Process

A. _____

B. _____

C. _____

D. _____

E. _____

F. _____

G. _____

H. _____

I. _____

J. _____

K. _____

L. _____

M. _____

N. _____

O. _____

P. _____

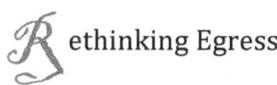

Name: _____

14

Create Alternatives, or Elements of Alternatives, Using Each Egress Process Component

A. _____

B. _____

C. _____

D. _____

E. _____

F. _____

G. _____

H. _____

I. _____

J. _____

K. _____

L. _____

M. _____

N. _____

O. _____

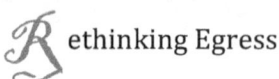

Name: _____

Create Alternatives, or Elements of Alternatives, Using Each Firefighting Process Component

A. _____

B. _____

C. _____

D. _____

E. _____

F. _____

G. _____

H. _____

I. _____

J. _____

K. _____

L. _____

M. _____

N. _____

O. _____

P. _____

Group Names: _____ **16**

Create and Embellish Alternatives, or Elements of Alternatives, Using Each Egress Process Component

A. _____

B. _____

C. _____

D. _____

E. _____

F. _____

G. _____

H. _____

I. _____

J. _____

K. _____

L. _____

M. _____

N. _____

O. _____

P. _____

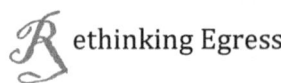

Group Names: _____ **17**

Create and Embellish Alternatives, or Elements of Alternatives, Using Each Firefighting Process Component

A. _____

B. _____

C. _____

D. _____

E. _____

F. _____

G. _____

H. _____

I. _____

J. _____

K. _____

L. _____

M. _____

N. _____

O. _____

P. _____

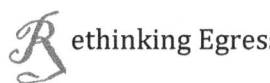

Name: _____

18

Evaluate Each of the Alternatives, or Elements of Alternatives, on Three Distinct Criteria:

> **Quality:** How good is the alternative IF it could be effectively implemented? Use a 1 to 9 evaluation scale where 1 is poor and 9 is great.

> **Feasibility:** What is the chance that the alternative could be effectively implemented within 10 years? Use a percentage evaluation scale, where 0 means there is no chance, 50 means there is a half chance, and 100 means that the alternative could surely be implemented.

> **Creativity:** How creative or innovative is the alternative? Use a 1 to 9 evaluation scale where 1 is a standard well-known idea and 9 is an idea that you had not heard of before.

Quality	Feasibility	Creativity	Alternative
(1 to 9)	(percent)	(1 to 9)	
____	____	____	_____
____	____	____	_____
____	____	____	_____
____	____	____	_____
____	____	____	_____
____	____	____	_____
____	____	____	_____
____	____	____	_____
____	____	____	_____
____	____	____	_____
____	____	____	_____

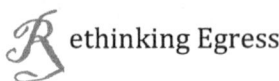

APPENDIX D: TECHNIQUES TO STIMULATE RECOGNITION OF VALUES

Use a wish list:
> If you had no constraints or limitations, what would you value?
> What elements constitute the "bottom line"?

Consider pros and cons of alternatives:
> Think of a great alternative. Why is it great?
> Think of a bad alternative. Why is it bad?
> Think of any alternative. What is good about it? Bad about it?
> Now use hypothetical alternatives with the above questions.

Imagine possible consequences:
> Think of the current situation. What is wrong with it? Right with it?
> Are there any unacceptable consequences?
> Are there any types of consequences that worry you?
> Are there possible consequences that you don't know how to control?

Use different perspectives:
> What would occupants of the building say matters?
> What would local politicians say mattered?
> What if the reason for egress was a biological release?

Use constraints and guidelines:
> Consider any constraints on the process of egress. Why is this of concern?
> Consider ant guidelines. What is their purpose?

Ask 'why' for each value:
> For each value on your list, consider why it matters. Their may be multiple answers each suggests another possible value. Repeat the process until you are not generating additional values.

www.ingramcontent.com/pod-product-compliance
Lightning Source LLC
Chambersburg PA
CBHW081849170526

45167CB00007B/2941